Zip Model of Dark Matter

Functionality

Zip Model of Dark Matter Functionality

S.B.Asoka Dissanayake

Asokaplus

Contents

Chapter 01

Introduction

Why Zip Model?

In my book on "Beyond the Known Horizon" I developed a theory to make wholesome look at the matter real (Visible Matter) and virtual matter (Dark Matter), taken together and not as individual entities. In reality they exist in tandem (together and in conjunction) and not in isolation. In that process, I briefly described the Zip Model to explain the manifestation of Dark Matter. I did not delve deep into depths, especially in regard to the formation and evolution of galaxies.

This book would do justice to both my theory and Zip Model that tries to describe the mechanics of dark matter.

There are reason for using the Zip Model.

The zip can be in both zipped or unzipped mode.

When zipped it gives stability.

When unzipped its gives it functionality.

I want to take dark matter out of the docile domain and put it in the right place in our universe.

The zipping structural arrangement does not interfere with its properties.

It keeps it fluid nature and let matter to pass through it.

It allows both expansion of space and given appropriate conditions contraction of space.

When I say contraction what it means is that the dark matter is transforming into matter real.

This contraction possibility is taken as model for galactic travel, a plausible trick with newly defined physics to make dark matter, a work horse of ingenuity.

Dark matter should give us the facility of slip sliding away in our galaxy and beyond.

I must write few lines about the zip fastener for completeness below. Like the cello tape it was one of the incredible simple but useful inventions of mankind which we fail to appreciate.

I am going to do justice to the Zip here but I wonder how could I pay tribute to the cello-tape.

Zip Fastener

A zipper, zip, fly or zip fastener, formerly known as a clasp locker, is a commonly used device for binding the edges of an opening of fabric or other flexible material, like on a garment or a bag. It is used in clothing (e.g., jackets and jeans), luggage and other bags, sporting goods, camping gear (e.g. tents and sleeping bags).

Whitcomb L. Judson was an American inventor from Chicago who was the first to conceive the idea, and to construct a workable zipper with interlocking teeth.

Initially it was called the "hookless fastener".

When you look closely at a zip fastener, you can see the intricacy of the two sides, with the dome or cup on each side with protruding teeth, which are locked or unlocked when the tab slider moves over them in one direction or the other. Something incredibly simple.

Advancements in machine tooling enabled precision cutting of the metal which meant that the zip fasteners were made accurately with consistent quality and both sides of the zipper being punched out of a single piece of wire.

The zipper is, of course, not without its own dangers, especially if fastened hastily over bare skin.

Imagine prepuce of the penis is caught.

I have seen few cases in my time with unbearable pain. This was demonstrated painfully in the popular 1998 Hollywood film "There's Something about Mary" and perhaps this more than anything else caused many men to seek out trousers without a zipper but a button fly.

For most of us, however, the zipper's helpfulness remains one of life's little overlooked pleasures.

Chapter 02

Zipping and Unzipping Model gives Flexibility and Modularity to Dark Matter?

Modularity is the degree to which a system's components may be separated and recombined.

The zippers has this property.

Zipping model can have many dimensions.

It can exist in a string formation or as a wave (vibrations can be in many ways). I prefer this wave to be like the vibration in a musical instrument with strings. The dark matter has no material form but it has the energy of its own to vibrate in many dimensions, since it does not have the structural or material hindrances.

When applied to the dark matter, the zipped formation can have a globular or twisted arrangement as seen in the globular galaxies and spiral galaxies.

Because of its open ends the ZIP can form network of connections, some very active some docile.

What is more pertinent is that the mathematics of the conformations can be constructed.

In addition the modularity theorem can be used to establish the connection between elliptic curves and modular forms.

The light warping around a dark matter into arcs and mirror images can be explained by describing the dark matter in zipped and or unzipped arrangement at a particular point in space in relation to matter.

Above all it can fit in with my hypothesis of rearrangement of dark matter to form matter and matter to form dark matter in their transformations.

Lot of things fall into place and modified string hypothesis has a place in the Zipped Model.

Only way we can figure out the presence of the dark matter is its relationship with the matter visible.

The gaps between the visible matter in the galaxies is not empty.

There are hot and cold gases and dirt particles.

It we can map these intermediates (with infrared and visible light) interfering with the dark matter, then we can have a rough sketch of the conformation of the virtual particles, the dark matter, in space.

It is a very difficult preposition.

Theoretically it is similar to the light sensitive and light insensitive particles of a photographic material. It is like the negative and positive photographic paper.

Mind you even negative though transparent is made of material.

Dark matter has no physical properties of that kind.

Chapter 03

Is it possible to explain the arrangement of the galaxies using zip model of dark matter?

Dark matter cannot be seen and has to be visualized as virtual particles.

What we can measure is either expansion of space (red shift) or contraction of space (blue shift).

By visualizing red shift and blue shift and by superimposition of matter (stars and galaxies) in space, the assembly of the dark matter has to be visualized in theory.

Then we can work out the possible interactions.

For this to manifest we have to postulate matter can be transformed into dark matter and dark matter to matter.

That is why I maintain they are two sides of the same coin.

Let us visualize the dark matter arrangement.

Like matter attract like matter and they merge and mix like water.

The force behind is the Dark Force that make virtual dark particles move about freely in space and also expand. Unlike matter that has to compete for its space forming atoms of increasing heavy mass (from hydrogen to helium to uranium) and density the dark matter has no hindrance whatsoever once formed.

It expands to the seamlessly space.

That is why there is abundance of dark matter and its existence is not governed by typical or accepted current physics.

The zipped dark matter may be thin or may be flat.

It can be coiled.

It can be a coiled coil.

I defer using a coiled coil, like in DNA since mechanics there is physical bonding between unlike particles of different charges reacting with each other while completing for limited space at subatomic levels.

Dark matter has no space hindrance.

In fact it owns the space.

It can be an opened zip at one end with variable length for each limb at one end and the next line of continuity may be a stretch (a gap in between the matter in reality) of tight zip and then the dark matter may open into (something impossible to scale) an unzipped dimension of considerable enormity and the next portion could be zipped again and forming imaginary loops that look like coils in material world that lets the matter (stars and galaxies) to align and rotate in one direction.

This would give birth to a globular galaxy of high density. The dark matter in the globular galaxy may end up in a dark holes or it may spring open with open arms of zip arrangement, of one, two or multiple ends.

The amount of dark matter is less and formation of matter is very limited in a globular galaxy.

Some of these globular forms may collapse so much due to their own gravity and they eventually end up as dark holes.

The unzipped portion would give the high density Starburst or irregular galaxy assuming that the unzipped portion of dark matter would change to matter (hydrogen to begin with) and collapse in space causing the galaxy to become smaller.

The scarcity of the dark matter probably would allow some of the larger galaxies to split off to smaller galaxies with intensely burning stars.

I am at a bit of a fix to explain the spiral galaxies.

Their beautiful shape is probably due to the speed of their rotation.

They are moving at tremendously high velocity and most likely due to the lack of dark matter stars at the periphery try to defy the gravity of the formation and escape to become independent galaxies.

Spiral arms are born as a random event.

Barred portion in the spiral galaxies are probably formed to retard the process of escaping of stars from the globular center and breaking off completely as new galaxies.

In that sense dark matter gives the stability.

In the same stretch of argument barred galaxies are old whereas the spiral galaxies without bars are younger.

How I am going to fit my ZIP theory?

Let me begin with the easy point of interpretation.

In the barred spiral at the point of the barred section, the matter is converted to dark matter to give its stability. Apart from that assumption how it is formed at the origin of the spiral is difficult to explain. It may be due to the speed of the rotation that matter is under tremendous stress and torn apart to subatomic particles and realigned to form dark matter with dark force.

It is feasible but what is the role of dark matter in the spiral without bars?

There is a halo between the disk and the spiral which is where the mysterious dark matter is resident.

We know it gives the stability to the disk shape of the galaxy.

Does the dark matter takes a neutral role, not changing into matter and hydrogen stars at all?

This is the contention, I wish I could dispel with my current hypothesis.

Chapter 04

Why I Love the Zip Model

I can take the "Big Bang" completely out in my model.

In the Zip Model there is no place for its existence.

It gives credence to the dark matter.

It integrates with matter real.

It can have seamless configurations.

Those are the plus points.

There are negative points of interest.

It does not explain why galaxies collide.

It is possible that it can minimize the car wreck damage the colliding galaxies would ensue.

We are witnessing such an event in the distant universe.

When it happens we would be able to better explain the mechanics behind the catastrophe.

Zip Model by nature is inherently simple.

Basically it can have two functional forms.

Zipped or unzipped.

It can model the string theory in practice but without many mathematical dimensions of string theory.

It makes way for a simple string theory.

It makes string theory plausible.

Zip Model can be used to make a simple model of dark matter (which we will never be able to see in reality).

Above all the zip theory gives randomness to galactic events.

The slider that zip and unzip is the Dark Force that inherently reside with the dark matter.

Chapter 05

This is my alternative theory to counteract the "Big Bang"

I have said, I am uncomfortable with the "Big Bang" and if I do not propose an alternative theory, then my scientific reasoning becomes VOID.

I want to propose a random theory not a unifying one.

It may sound bizarre but it is tangible in theory.

The space is taken into account with its virtual particles.

In this theory of "Whole of the Universe" is taken as one conglomerate.

The matter real is only the small change.

It is less than 5%.

The dark matter is said to be 25%.

The rest 70% (near enough) is dark energy.

My assumption is that matter which carry energy with charge would always move towards a more stable state without charge.

Only way for it to make that move is to become dark matter and in that process it releases dark energy and dark particles without charge.

This in fact, makes the space expand.

The matter moving out and universe expanding are two sides of the coin.

It is the virtual reality.

In this theory the matter reduces not by constant shift but by random events.

So when the matter is reduced to a critical point, say less than 1% something intangible has to happen.

There is a mismatch.

At this point Dark energy and Dark Matter cannot rule the universe.

It has to go into reverse gear, not in a "Big Bang" but in a series of steps.

So the dark energy and dark matter is consumed and new matter is formed and the worlds, galaxies and new universes are formed.

There is neither beginning nor end, only a process, somewhat akin to cyclical phenomena (forward-dark matter and backwards-transformation to matter) are in existence.

Better way to visualize is zipping and unzipping of a fastener.

The term cyclical is a misnomer for this transformation of energy particles.

It falls into quantum dynamics, not a cycle, like day and night.

It is not like birth or bust or bubble.

There is only two or three dimensions to this theory but if few more dimensions are mathematically extrapolated, the possibilities are infinite.

Dark matter is docile in its very nature.

It tend to keep company with likes of it.

Since it has no electromagnetic forces or charges within its particles, there is no wonder they keep company.

What is keeping them intact without wandering away?

So we have to postulate there is dark force keeping them together in its own gravitation sphere.

This gravitational force is different from the gravitational forces binding matter.

In fact, it has anti-gravitational power and makes the universe expand.

In this scenario space expands and the matter real (like a pictures drawn on the surface of a balloon) all around, gets pulled apart together.

Even though, it portent to be docile, it is not.

It makes a real contribution to the existence and the functionality of the matter.

I go even further and postulate that dark matter changes to matter given the right circumstances.

That of course is the core material in this book.

Suppose a bit of dark material in its solo shell come into the vicinity of a galaxy which has enough and more dark matter to pull them towards.

What is the likely outcome.

Simple explanation is for it to unite with the likes.

Nothing else but expanding space within the galaxy..

The other scenario is to strip the galaxy of its dark matter.

The galaxy has to contract within and the dark matter finds its own space in the universe.

But what if this process is different altogether, simply because of its docile nature.

The two dark bodies would come into near enough contact range but without merging but wandering about happily and aimlessly.

What tilts the balance is not dark matter but matter itself.

For instance, the material in a star is burning intensely and spinning violently.

By random event, it strips a string out of the dark matter on its outer periphery,from the solo unit..

The dark mater spins into a different mode by this action.

It starts forming matter and start striping little bit more from the dark matter.

What starts as a random event now blows out of proportion to become a hydrogen star.

A star is born.

Not one may be more!

The simplest of atom, the hydrogen is the outcome when matter is born out of dark matter transformation.

The way to test this theory is to test whether matter can be changed to dark matter or dark forces.

This is more feasible since only charged particles are necessary to neutralize the charged particles of matter and certainly not antimatter.

No annihilation is envisaged.

The converse of it is to try change dark matter to matter.

Bombarding dark matter to make matter is assumed more difficult due to its stability and its inherent expansion.

In other words contracting (compacting or concentrating space) the space is insurmountable.

If that can be achieved "the time travel" actually becomes feasible.

Chapter 06

Does The Zip Model points to a Stable Universe?

I would rather not wish that the Zip Model should give a stable structure to the dark matter.

But if the universe is stable the zipping and unzipping should stop at a particular point in its evolution.

Then what?

The galaxies would collapse by their own gravity.

The universe stops expanding.

Only the visible matter changes on its own accord without any interaction with the dark matter.

In other words the matter overrides the influence of dark matter.

All these are improbable and I would like to reject a stable universe.

Chapter 07

Does The Zip Model points to an Unstable Universe?

There is more than one reason to accept an unstable universe.

It fits in nicely with the zipped and unzipped modes, giving rise to many formations and calamity.

Both expansion and colliding galaxies are plausible.

Then we need to time these mega events.

That adds an unnecessary complication to the Zip Model.

At what point should the time element be attached to the model?

Are we to go back to the "Big Bang" of beginning time Zero?

The easily explainable expansion in space albeit with some retardation and a predictable outcome in time frame makes the model attractive in theory.

Is it possible to fix a time frame to the visible matter but not to the dark matter?

The dilemma!

Which came first?

The matter first?

Dark matter second?

Are they simultaneous?

Then why more dark matter?

An unstable universe with a time tagged will produce more questions than answers.

Chapter 08

Does The Zip Model points to an Evolving Universe?

No Beginning and No End

This is how I like to figure out the Zip Model.

The zipping and unzipping is a current event.

It is neither a zipped nor unzipped phenomenon in relation to the reality of time element.

What I mean is if one locates a zipped portion (meaning dark matter is docile form) one cannot finds its timing or the aging in relation to another piece which is in turmoil and fully (meaning lot of burning stars around) unzipped.

The zipped location can become unzipped and unzipped location can become zipped with minimum of disruption to the rest of the universe.

This also make the tendency for spacing of galaxies in the universe in such a fashion that impending collusion is not inevitable but a plausible probability.

The interaction of dark matter within the boundaries of the Zip Model may either retard or expand the space in any dimension (akin to string theory of many dimension but limited in scope - not the endless dimensions that can be worked out in mathematics) but seems neither past nor future to the observer.

The dark matter can warp time and space.

In other words contracting the space would mean we are advancing towards a nearby galaxy (blue shift instead red shift).

The universe becomes a football field in this dimension.

The new physics is the ball in play!

We need to go for an alternative physics.

The Zip Model is one of the many possibilities of the evolving universe.

Chapter 09

Are there any other possibilities?

The bone of contention here is why are we having only one hypothesis, the "Big Bang"?

The same argument can be raised against the theory I have postulated.

Zip Model is one way of looking at it.

There could be other models.

What are they?

Do we need to consider time as a dimension like in the Big Bang Model?

Could we leave the time as a concept that we need to answer the question beginning of the universe.

What is point in (except to satisfy the church or religion one believes) of knowing it.

By knowing the time of the beginning would us be able to create a new universe?

I have no problem with using light years as a measure and working the distance between astral objects.

I have no problem of using time for global travel.

The air industry needs a time table and time changes according to latitudes of this planet.

Time has enormous importance in dissecting our origin in our planet.

Beyond that it has no use in investigating the universe.

That is my point of view.

Especially investigating the dark matter.

Its beginning has no meaning to scientific inquiry.

We cannot work out the physics of dark matter with or without time.

If an earthling travels out at the speed of light and if the earthling return after 50 years later, the earthlings on earth seem to be much older for the universe bound traveling earthling.

The Zip Model is worked out disregarding the time concept.

It gives room for thinking afresh.

Similarly there can be other models.

Say, for instance water or fluid model for dark matter.

Something that flows and permeate the universe.

There can be models for observed expansion of the universe which is different from the "Big Bang".

What are they?

I leave the reader to escape from the bondage with Big Bang.

This book is to stir up the investigating mind of a natural scientist or a budding physicist.

If I have achieved that, I consider this book is a success.

Authors Note

This is the feeling one gets when one is on a space odyssey

Zip Sliding Away

David Bowie - Space Oddity

> Ground Control to Major Tom
> Ground Control to Major Tom
> Take your protein pills
> and put your helmet on
>
> Ground Control to Major Tom
> Commencing countdown,
> engines on
> Check ignition
> and may God's love be with you
>
> Ten, Nine, Eight, Seven, Six, Five, Four, Three, Two, One,
> Liftoff
>
> This is Ground Control
> to Major Tom
> You've really made the grade
> And the papers want to know whose shirts you wear
> Now it's time to leave the capsule, if you dare

This is Major Tom to Ground Control

I'm stepping through the door

And I'm floating

in a most peculiar way

And the stars look very different today

For here

Am I sitting in a tin can

Far above the world

Planet Earth is blue

And there's nothing I can do

Though I'm past

one hundred thousand miles

I'm feeling very still

And I think my spaceship knows which way to go

Tell my wife I love her very much

she knows

Ground Control to Major Tom

Your circuit's dead,

there's something wrong

Can you hear me, Major Tom?

Can you hear me, Major Tom?

Can you hear me, Major Tom?

Can you....

Here am I floating
round my tin can
Far above the Moon
Planet Earth is blue
And there's nothing I can do.

Paul Simon - Slip Sliding Away

Slip sliding away
Slip sliding away
You know the nearer your destination
The more you're slip sliding away

I know a man
He came from my home town
He wore his passion for his woman
Like a thorny crown
He said Dolores
I live in fear
My love for you is so overpowering
I'm afraid that I will disappear

Slip sliding away
Slip sliding away
You know the nearer your destination
The more you're slip sliding away

I know a woman
Became a wife
These are the very words she uses
To describe her life
She said a good day

Ain't got no rain

She said a bad day's when I lie in bed

And think of things that might have been

Slip sliding away

Slip sliding away

You know the nearer your destination

The more you're slip sliding away

And I know a father

Who had a son

He longed to tell him all the reasons

For the things he'd done

He came a long way

Just to explain

He kissed his boy as he lay sleeping

Then he turned around and headed home again

Slip sliding away

Slip sliding away

You know the nearer your destination

The more you're slip sliding away

God only knows

God makes his plan

The information's unavailable

To the mortal man
We're working our jobs
Collect our pay
Believe we're gliding down the highway
When in fact we're slip sliding away

Slip sliding away
Slip sliding away
You know the nearer your destination
The more you're slip sliding away

Slip sliding away
You know the nearer your destination
The more you're slip sliding away
Mmm...

Asokaplus

www.ingramcontent.com/pod-product-compliance
Lightning Source LLC
Chambersburg PA
CBHW080609190526
45169CB00007B/2947